AI 写实人物
绘画关键词图鉴 Stable Diffusion 版

AIGC-RY 研究所 著

人民邮电出版社

北京

图书在版编目（ＣＩＰ）数据

AI写实人物绘画关键词图鉴：Stable Diffusion版 / AIGC-RY研究所著. -- 北京：人民邮电出版社，2023.10
ISBN 978-7-115-62565-6

Ⅰ．①A… Ⅱ．①A… Ⅲ．①图像处理软件 Ⅳ．
①TP391.413

中国国家版本馆CIP数据核字(2023)第175718号

内 容 提 要

AI是当下无法阻挡的艺术创作趋势。

本书首先简要地介绍了关键词的使用方法，以及软件界面的参数，帮助读者大致了解生成图片的基本原理；正文实战部分，展示了画质渲染、构图视角、画面效果、人物类型、头发样式、头部细节、服装风格、人物动作、场景呈现等9大主题的AI图片生成效果，并给出了提示词说明，通过图文对应的方式帮助读者了解生成图片的具体方法，从而生成自己想要的图像。

本书适合对AI图像创作感兴趣的读者和有AI图像创作需求的设计师、插画师等阅读。

◆ 著　　　AIGC-RY 研究所
　责任编辑　王　铁
　责任印制　周昇亮

◆ 人民邮电出版社出版发行　北京市丰台区成寿寺路 11 号
　邮编　100164　电子邮件　315@ptpress.com.cn
　网址　　https://www.ptpress.com.cn
　北京捷迅佳彩印刷有限公司印刷

◆ 开本：700×1000　1/16
　印张：9　　　　　　　　　　　2023 年 10 月第 1 版
　字数：197 千字　　　　　　　2024 年 9 月北京第 2 次印刷

定价：49.80 元

读者服务热线：(010)81055296　印装质量热线：(010)81055316
反盗版热线：(010)81055315
广告经营许可证：京东市监广登字 20170147 号

关键词的使用方法

关键词基础模板

使用英文单词更有效,单词之间使用英文半角状态下的逗号(,)作为间隔,逗号前后带不带空格不影响关键词输入,出图整体风格与使用的模型库相关(偏照片、写实或动漫风格)。

基础的关键词模板由以下几部分组成。

● 关键词分隔

使用半角状态下的英文逗号,可以分隔不同的关键词标签。除此之外,空格和换行等不影响标签分隔。

● 正向提示词参考

4k,best quality,masterpiece   画质:用于提升或保持画面的整体质量

full body,face to camera,close shot   主体内容:描述画面的主体部分

casual wear,smile   附加内容:与画面主体有关的设定,可以让画面主体内容更加完善和丰富

● 反向提示词参考

NSFW   不良信息:过滤敏感信息

bad anatomy,bad hands fingers   主体内容:对整体质量与画面主体内容相关的反向提示

low quality   语义失衡:避免出图产生额外干扰信息

Stable Diffusion 使用方法

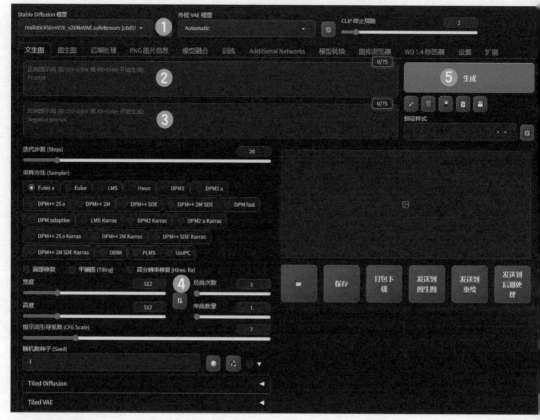

❶ 模型库
根据需求选择相应风格的模型库。

❷ 提示词
用文字（英文单词）描述想生成的画面内容。

❸ 反向词
用文字（英文单词）描述不想生成的画面内容。

❹ 图片编辑
调整生成图片的大小、批次和数量。

❺ 生成图片
点击"生成"按钮即可得到生成的图片。

目录 CONTENTS

Chapter 01	画质渲染	006
Chapter 02	构图视角	012
Chapter 03	画面效果	020
Chapter 04	人物类型	034
Chapter 05	头发样式	052
Chapter 06	头部细节	062
Chapter 07	服装风格	080
Chapter 08	人物动作	104
Chapter 09	场景呈现	120

Chapter 01

画质渲染

画面质量是指图像或视频呈现的视觉效果的好坏程度，它与图像的分辨率、清晰度、细节、色彩准确性和整体表现等有关。本章将介绍一些常见的与画面质量相关的关键词。

高分辨率 high resolution

提示词：**high resolution**,best quality,masterpiece,upper_body,face to camera,casual wear,smile

极其详细的刻画 extreme detail description

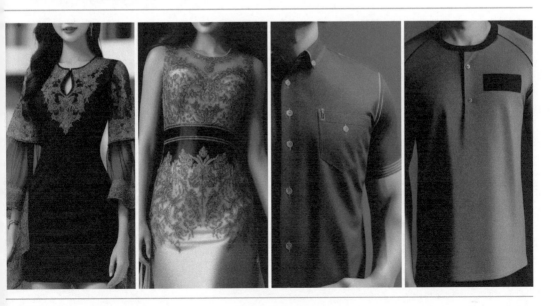

提示词：**extreme detail description**,best quality,masterpiece

● **虚幻引擎** unreal engine

提示词：**unreal Engine**,best quality,masterpiece,1 girl solo,face to camera,casual wear,smile

● **V 射线** v-ray

提示词：**V-ray**,best quality,masterpiece, upper_body, face to camera,1 girl solo,casual wear,smile

● OC 渲染　octane rendering

提示词：**octane rending**,best quality,masterpiece

● 渲染　maxon cinema 4D

提示词：**maxon cinema 4D**,best quality,masterpiece, upper_body, face to camera,smile

● 阴影效果渲染 shadow effect rendering

提示词：**Shadow effect rending**,best quality,masterpiece

● 真实感渲染 quixel megascans rendering

提示词：**quixel megascans rendering**,best quality,masterpiece

AI绘画启蒙

3天快速上手Stable Diffusion

开启AI绘画的 ∞ 可能

3步掌握
1. 认识 Stable Diffusion
2. 熟悉软件界面
3. 一键生成图片

扫码添加客服
回复【162565】

免费
领取课程

建筑渲染　architectural rendering

提示词：**architectural rendering**,best quality,masterpiece

室内渲染　indoor rendering

提示词：**indoor rendering**,best quality,masterpiece

Chapter 02

构图视角

构图视角是指在绘画、摄影和其他可视艺术形式中，艺术家或摄影师选择和安排元素的方式，用以创造出具有美感和视觉吸引力的图像。构图视角涉及观察者的位置和角度，以及在画面或图像中放置和组织元素的方式，它会影响观者对图像的感知、情感和关注的焦点。

全身像 full shot

提示词：4k,best quality,masterpiece,full body,face to camera,full shot,smile

上半身 upper_body

提示词：4k,best quality,masterpiece,face to camera,upper_body,smile

● 近景 close shot

提示词：4k,best quality,masterpiece,upper_body,face to camera,close shot, casual wear,smile

● 自拍 selfie

提示词：4k,best quality,masterpiece,upper_body,face to camera,selfie,1 girl solo,casual wear,smile

● 特写 close-up

提示词：4k,best quality,masterpiece,upper_body,face to camera,close-up, casual wear,smile

● 微距摄像 macro shot

提示词：4k,best quality,masterpiece,face to camera,macro shot,1 girl solo,smile

● 顶视图 top view

提示词：4k,best quality,masterpiece,full body,face to camera,top view, casual wear,smile

● 侧视图 side view

提示词：4k,best quality,masterpiece,upper_body,face to camera,side view,1 girl solo,smile

● 聚焦在面部　face_focus

提示词：4k,best quality,masterpiece,portrait,face to camera,face_focus, 1 girl solo,casual wear,smile

● 聚焦在眼睛上　eyes_focus

提示词：4k,best quality,masterpiece,portrait,face to camera,eyes_focus,casual wear,smile

● 聚焦在单个人物上　solo_focus

提示词：4k,best quality,masterpiece,portrait,face to camera,solo_focus, 2 girls,casual wear,smile

● 后背视角　from_back

提示词：4k,best quality,masterpiece,full body,from_back,casual wear

倾斜角度　dutch_angle

提示词：4k,best quality,masterpiece,face to camera,dutch_angle,casual wear,smile

低角度　low_angle

提示词：4k,best quality,masterpiece,full body,face to camera,low_angle,1 girl solo,casual wear,smile

Chapter 03
画面效果

画面效果是指在电影、电视、游戏或其他媒体中,使用计算机图形技术和特殊效果创造出的视觉元素和场景。画面效果旨在增强观者的视觉体验,创造出通过传统拍摄手法无法实现的效果,例如各种光照和不同风格的画面呈现。

正面光 frontlight

提示词：4k,best quality,masterpiece,upper_body,face to camera, **frontlight**,casual wear,smile

侧面光 sidelight

提示词：4k,best quality,masterpiece,upper_body,face to camera, **sidelight**,casual wear,smile

● 逆光 backlight

提示词：4k,best quality,masterpiece,upper_body,face to camera, **backlight**,1 girl solo,smile

● 边缘光 rim_light

提示词：4k,best quality,masterpiece,upper_body,face to camera, **rim_light**,casual wear,smile

环境光 ambient_light

提示词：4k,best quality,masterpiece,upper_body,face to camera, ambient_light,casual wear,smile

阳光 sunlight

提示词：4k,best quality,masterpiece,upper_body,face to camera, sunlight,casual wear,smile

● **斑驳的阳光** dappled_sunlight

提示词：4k,best quality,masterpiece,upper_body,face to camera,dappled_sunlight,1 girl solo,casual wear,smile

● **月光** moonlight

提示词：4k,best quality,masterpiece,face to camera,moonlight,smile

霓虹灯 neon

提示词：4k,best quality,masterpiece,upper_body,face to camera,neon,casual wear,smile

自然光 natural light

提示词：4k,best quality,masterpiece,upper_body,face to camera,natural light,1 girl solo,casual wear,smile

● 游戏 CG game_cg

提示词：4k,best quality,masterpiece,upper_body,face to camera,**game_cg**,smile

● 写真 photo_album

提示词：4k,best quality,masterpiece,upper_body,face to camera,**photo_album**,girl,casual wear,smile

● 封面 cover_page

提示词：4k,best quality,masterpiece,upper_body,face to camera,**cover_page**,girl,casual wear,smile

● 复古风格 retro_style

提示词：4k,best quality,masterpiece,upper_body,face to camera,**retro_style**,girl,casual wear,smile

● **杂志封面** magazine_cover

提示词：4k,best quality,masterpiece,upper_body,face to camera,magazine_cover,girl,casual wear,smile

● **海报** poster

提示词：4k,best quality,masterpiece,upper_body,face to camera,poster,casual wear,smile

写实 realistic

提示词：4k,best quality,masterpiece,upper_body,face to camera,**realistic**,casual wear,smile

照片风格 photo_medium

提示词：4k,best quality,masterpiece,upper_body,face to camera,**photo_medium**,casual wear,smile

● 油画 oil_painting

提示词：4k,best quality,masterpiece,face to camera,oil_painting,girl,smile

● 拍立得风格 polaroid

提示词：4k,best quality,masterpiece,upper_body,face to camera,polaroid,casual wear,smile

三格 3koma

提示词：4k,best quality,masterpiece,upper_body,face to camera,3koma,girl,casual wear,smile

柔和的色彩 pastel_color

提示词：4k,best quality,masterpiece,portrait,face to camera,pastel_color,casual wear,smile

● 素描　sketch_pencil

提示词：4k,best quality,masterpiece,portrait,face to camera,sketch_pencil,girl,casual wear,smile

● 彩色铅笔　colored_pencil

提示词：4k,best quality,masterpiece,portrait,face to camera,colored_pencil,girl,casual wear,smile

水彩 watercolor

提示词：4k,best quality,masterpiece,portrait,face to camera,**watercolor**,girl,casual wear,smile

丙烯颜料画 acrylic_paint

提示词：4k,best quality,masterpiece,portrait,face to camera,**acrylic_paint**,casual wear,smile

Chapter 04
人物类型

通常根据某种分类标准将人物分为不同的类型,这些分类标准包括性别、年龄、关系、属性等。人物类型可以帮助我们更好地理解和描绘不同人群的特征。

女人 female

提示词：4k,best quality,masterpiece,face to camera,female,casual wear,smile

男人 male

提示词：4k,best quality,masterpiece,upper_body,face to camera,male,smile

● 2个女孩 2 girls

提示词：4k,best quality,masterpiece,full body,face to camera,**2 girls**,smile

● 2个男孩 2 boys

提示词：4k,best quality,masterpiece,face to camera,**2 boys**,smile

幼童 toddler

提示词：4k,best quality,masterpiece,full body,face to camera,**toddler**,casual wear,smile

儿童 child

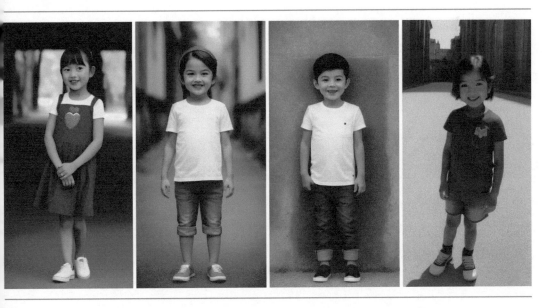

提示词：4k,best quality,masterpiece,full body,face to camera,**child**,casual wear,smile

● 青年 teenage

提示词：4k,best quality,masterpiece,full body,face to camera,teenage,school uniform,smile

● 老年 old

提示词：4k,best quality,masterpiece,upper_body,face to camera,old, casual wear,smile

大叔 bara

提示词：4k,best quality,masterpiece,face to camera,bara,casual wear,smile

成熟女性 mature_female

提示词：4k,best quality,masterpiece,face to camera,mature_female,casual wear,smile

● 姐妹 sisters

提示词：4k,best quality,masterpiece,face to camera,**sisters**,casual wear,smile

● 兄弟姐妹 siblings

提示词：4k,best quality,masterpiece,face to camera,**siblings**,smile

母子 mother and son

提示词：4k,best quality,masterpiece,full body,face to camera,mother and son,casual wear,smile

母女 mother and daughter

提示词：4k,best quality,masterpiece,face to camera,mother and daughter,casual wear,smile

● 夫妻 husband and wife

提示词：4k,best quality,masterpiece,face to camera,**husband and wife**,smile

● 萝莉 loli

提示词：4k,best quality,masterpiece,face to camera,**loli**,smile,1 girl solo

正太 shota

提示词：4k,best quality,masterpiece,face to camera,**shota**,casual wear,smile,1 boy

辣妹 gyaru

提示词：4k,best quality,masterpiece,full body,face to camera,**gyaru**,smile,1 girl solo

● 丰满 plump

提示词：4k,best quality,masterpiece,face to camera,**plump**,1 girl solo,smile

● 肥胖 fat

提示词：4k,best quality,masterpiece,face to camera,**fat**,smile

怀孕　pregnant

提示词：4k,best quality,masterpiece,face to camera,**pregnant**,1 girl solo,casual wear,smile

肌肉　muscular

提示词：4k,best quality,masterpiece,face to camera,**muscular**,smile

白皙皮肤 fair skin

提示词：4k,best quality,masterpiece,face to camera,fair skin,1 girl solo,casual wear,smile

小麦色皮肤 wheat colored skin

提示词：4k,best quality,masterpiece,face to camera,wheat colored skin,smile

胸肌 pectorals

提示词：4k,best quality,masterpiece,upper_body,**pectorals**,male

小胸部 small_chest

提示词：4k,best quality,masterpiece,upper_body,**small_chest**,1 girl solo

膝盖 knee

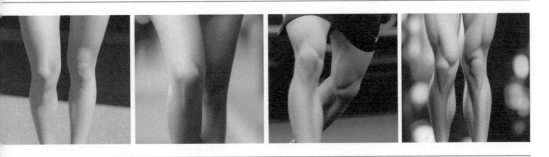

提示词：4k,best quality,masterpiece,lower_body,**knee**

● 后背曲线 back curve

提示词：4k,best quality,masterpiece,upper_body,**back curve**,1 girl solo

● 裸双肩 bare_shoulders

提示词：4k,best quality,masterpiece,upper_body,**bare_shoulders**,1 girl solo

● 锁骨 collarbone

提示词：4k,best quality,masterpiece,upper_body,**collarbone**,1 girl solo

腋窝 armpits

提示词：4k,best quality,masterpiece,upper_body,**armpits**,1 girl solo

腰 waist

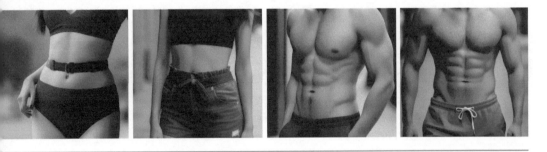

提示词：4k,best quality,masterpiece,upper_body,**waist**

肚子 belly

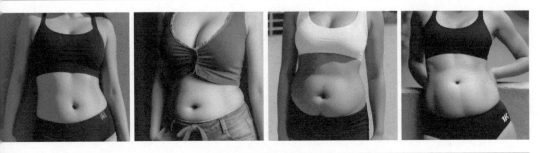

提示词：4k,best quality,masterpiece,upper_body,**belly**,1 girl solo

腹肌 abs

提示词：4k,best quality,masterpiece,upper_body,**abs**

肋骨 ribs

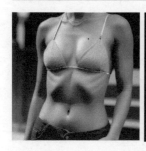   

提示词：4k,best quality,masterpiece,upper_body,**ribs**,1 girl solo

腹部 midriff

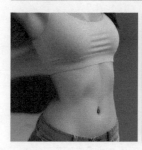   

提示词：4k,best quality,masterpiece,upper_body,**midriff**,1 girl solo

胯部 crotch

提示词：4k,best quality,masterpiece,upper_body,**crotch**

宽胯 wide_hips

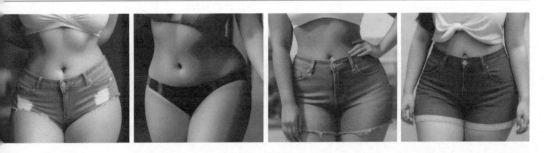

提示词：4k,best quality,masterpiece,upper_body,**wide_hips**,1 girl solo

臀部 buttock

提示词：4k,best quality,masterpiece,**buttock**

Chapter 05

头发样式

可以根据个人喜好、时尚潮流、文化背景、职业需求等因素来选择和设计不同的头发样式。本章是一些常见的头发样式的展示。

● 直发 straight_hair

提示词：4k,best quality,masterpiece,portrait, face to camera,**straight_hair**,1 girl solo,casual wear,smile

● 卷发 curly_hair

提示词：4k,best quality,masterpiece,portrait, face to camera,**curly_hair**,1 girl solo,casual wear,smile

● 波波头 bob_cut

提示词：4k,best quality,masterpiece,portrait, face to camera,**bob_cut**,1 girl solo,casual wear,smile

● 公主头 princess_head

提示词：4k,best quality,masterpiece, portrait, face to camera,**princess_head**,1 girl solo, casual wear,smile

● **齐刘海** blunt_bangs

提示词：4k,best quality,masterpiece,portrait, face to camera,**blunt_bangs**,1 girl solo, casual wear,smile

● **马尾辫** ponytail

提示词：4k,best quality,masterpiece,portrait, **ponytail**,1 girl solo,casual wear,smile

● **低扎头发** low_tied_hair

提示词：4k,best quality,masterpiece,portrait, face to camera,**low_tied_hair**,1 girl solo, casual wear,smile

● **双马尾** twintails

提示词：4k,best quality,masterpiece, portrait, face to camera,**twintails**,1 girl solo, casual wear,smile

● 法式辫子 french_braid

提示词：4k,best quality,masterpiece,portrait, face to camera,**french_braid**,1 girl solo, casual wear,smile

● 双辫子 twin_braids

提示词：4k,best quality,masterpiece,portrait, face to camera,**twin_braids**,1 girl solo, casual wear,smile

● 侧辫子 side_braid

提示词：4k,best quality,masterpiece,portrait, face to camera,**side_braid**,1 girl solo, casual wear,smile

● 麻花辫马尾 braided_ponytail

提示词：4k,best quality,masterpiece,portrait, **braided_ponytail**,1 girl solo,casual wear,smile

● 双发髻 double_buns

提示词：4k,best quality,masterpiece,portrait, face to camera,**double_buns**,1 girl solo, casual wear,smile

● 丸子头 hair_bun

提示词：4k,best quality,masterpiece,portrait, face to camera,**hair_bun**,1 girl solo, casual wear,smile

● 芭蕾发髻 ballet_hair_bun

提示词：4k,best quality,masterpiece,portrait, face to camera,**ballet_hair_bun**,1 girl solo, casual wear,smile

● 波浪卷 wavy_hair

提示词：4k,best quality,masterpiece,portrait, face to camera,**wavy_hair**,1 girl solo, casual wear,smile

● 爆炸头 afro

提示词：4k,best quality,masterpiece,portrait, face to camera,**afro**,male,casual wear,smile

● 光头 bald

提示词：4k,best quality,masterpiece,portrait, face to camera,**bald**,male,casual wear,smile

● 长发 long_hair

提示词：4k,best quality,masterpiece,portrait, face to camera,**long_hair**,1 girl solo, casual wear,smile

● 短发 short_hair

提示词：4k,best quality,masterpiece,portrait, face to camera,**short_hair**,1 girl solo, casual wear,smile

● 中长发 medium_hair

提示词：4k,best quality,masterpiece,portrait,face to camera,**medium_hair**,1 girl solo,casual wear,smile

● 黑色头发 black_hair

提示词：4k,best quality,masterpiece,portrait,face to camera,**black_hair**,1 girl solo,casual wear,smile

● 蓝色头发 blue_hair

提示词：4k,best quality,masterpiece,portrait,face to camera,**blue_hair**,1 girl solo,casual wear,smile

● 绿色头发 green_hair

提示词：4k,best quality,masterpiece,portrait,face to camera,**green_hair**,1 girl solo,casual wear,smile

● 棕色头发 brown_hair

提示词：4k,best quality,masterpiece,portrait,face to camera,**brown_hair**,1 girl solo,casual wear,smile

● 粉色头发 pink_hair

提示词：4k,best quality,masterpiece,portrait,face to camera,**pink_hair**,1 girl solo,smile

● 红色头发 red_hair

提示词：4k,best quality,masterpiece,portrait,face to camera,**red_hair**,1 girl solo,smile

● 银发 silver_hair

提示词：4k,best quality,masterpiece,portrait,face to camera,**silver_hair**,1 girl solo,smile

● 灰白发 grey_hair

提示词：4k,best quality,masterpiece,portrait, face to camera,**grey_hair**,1 girl solo,casual wear,smile

● 金发 blonde_hair

提示词：4k,best quality,masterpiece,portrait, face to camera,**blonde_hair**,1 girl solo, casual wear,smile

● 凌乱的头发 messy_hair

提示词：4k,best quality,masterpiece,portrait, face to camera,**messy_hair**,1 girl solo, casual wear,smile

● 柔顺的头发 smooth_hair

提示词：4k,best quality,masterpiece,portrait, face to camera,**smooth_hair**,1 girl solo, casual wear,smile

蓬松的头发 fluffy_hair

提示词：4k,best quality,masterpiece,portrait, face to camera,**fluffy_hair**,1 girl solo, casual wear,smile

湿发 wet_hair

提示词：4k,best quality,masterpiece,portrait, face to camera,**wet_hair**,1 girl solo,smile

背头 slicked_back hair

提示词：4k,best quality,masterpiece,portrait, face to camera,**slicked_back hair**,male, casual wear,smile

拳击辫 boxing braid

提示词：4k,best quality,masterpiece,portrait, face to camera,**boxing braid**,1 girl solo, casual wear,smile

Chapter 06

头部细节

头部细节包括人脸上的微小特征和细节,例如表情、皱纹、肌肉运动、皮肤纹理等,以及头部的一些装饰,例如眼镜、帽子等,这些细节对于表达情感、传递信息和呈现真实感非常重要。

● 明亮的眼睛 light_eyes

提示词：4k,best quality,masterpiece,portrait, face to camera,**light_eyes**,1 girl solo, casual wear,smile

● 闭上眼睛 closed_eyes

提示词：4k,best quality,masterpiece,portrait, face to camera,**closed_eyes**,1 girl solo, casual wear,smile

● 蒙上眼睛 blindfolded_eyes

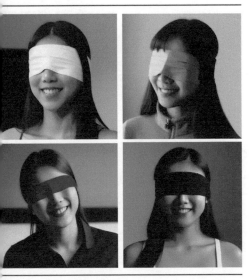

提示词：4k,best quality,masterpiece,portrait, face to camera,**blindfolded_eyes**,1 girl solo, smile

● 睁大眼睛 wide-eyed

提示词：4k,best quality,masterpiece, portrait, face to camera,**wide-eyed**,smile

● 黑色瞳孔 black_eyes

提示词：4k,best quality,masterpiece,portrait, face to camera,**black_eyes**,smile

● 绿色瞳孔 green_eyes

提示词：4k,best quality,masterpiece,portrait, face to camera,**green_eyes**,casual wear,smile

● 蓝色瞳孔 blue_eyes

提示词：4k,best quality,masterpiece,portrait, face to camera,**blue_eyes**,casual wear,smile

● 塌鼻梁 collapsed nose

提示词：4k,best quality,masterpiece, portrait, face to camera,**collapsed nose**,1 girl solo,smile

● 高鼻梁 high nose bridge

提示词：4k,best quality,masterpiece,portrait, face to camera,**high nose bridge**,male, casual wear,smile

● 精致的鼻子 delicate_nose

提示词：4k,best quality,masterpiece,portrait, face to camera,**delicate_nose**,1 girl solo, casual wear,smile

● 张嘴 open_mouth

提示词：4k,best quality,masterpiece,portrait, face to camera,**open_mouth**,smile

● 含奶嘴 pacifier

提示词：4k,best quality,masterpiece,portrait, face to camera,**pacifier**,1 girl solo

● 露出上牙 upper_teeth

提示词：4k,best quality,masterpiece,portrait,face to camera,**upper_teeth**,smile

● 闭嘴 close_mouth

提示词：4k,best quality,masterpiece,portrait,face to camera,**close_mouth**,casual wear

● 露出舌头 tongue

提示词：4k,best quality,masterpiece,portrait,face to camera,**tongue**,1 girl solo

● 抿嘴 pursed lips

提示词：4k,best quality,masterpiece,portrait,face to camera,**pursed lips**,1 girl solo,casual wear

● 项链　necklace

提示词：4k,best quality,masterpiece,portrait,
face to camera,**necklace**,1 girl solo,smile

● 化妆　makeup

提示词：4k,best quality,masterpiece,portrait,
face to camera,**makeup**,1 girl solo,smile

● 雀斑　freckles

提示词：4k,best quality,masterpiece,portrait,
face to camera,**freckles**,smile

● 小胡子　mustache

提示词：4k,best quality,masterpiece,portrait,
face to camera,**mustache**,male,casual
wear,smile

● 络腮胡 collateral beard

提示词：4k,best quality,masterpiece,portrait, face to camera,**collateral beard**,male, casual wear,smile

● 头盔 helmet

提示词：4k,best quality,masterpiece,portrait, face to camera,**helmet**,casual wear,smile

● 眼镜 glasses

提示词：4k,best quality,masterpiece,portrait, face to camera,**glasses**,casual wear,smile

● 贝雷帽 beret

提示词：4k,best quality,masterpiece,upper_ body,face to camera,**beret**,casual wear,smile

● 派对帽 party_hat

提示词：4k,best quality,masterpiece,upper_body,face to camera,**party_hat**,smile

● 高顶礼帽 top_hat

提示词：4k,best quality,masterpiece,upper_body,face to camera,**top_hat**,smile

● 学位帽 mortarboard

提示词：4k,best quality,masterpiece,portrait,face to camera,**mortarboard**,male,smile

● 海盗帽 pirate_hat

提示词：4k,best quality,masterpiece,portrait,face to camera,**pirate_hat**,casual wear,smile

● 渔夫帽　bucket_hat

提示词：4k,best quality,masterpiece,portrait,face to camera,**bucket_hat**,casual wear,smile

● 安全帽　hardhat

提示词：4k,best quality,masterpiece,portrait,face to camera,**hardhat**,casual wear,smile

● 草帽　straw_hat

提示词：4k,best quality,masterpiece,upper_body,face to camera,**straw_hat**,1 girl solo,smile

● 棒球帽　baseball_cap

提示词：4k,best quality,masterpiece,upper_body,face to camera,**baseball_cap**,smile

● 鸭舌帽 flat_cap

提示词：4k,best quality,masterpiece,portrait,face to camera,**flat_cap**,casual wear,smile

● 牛仔帽 cowboy_hat

提示词：4k,best quality,masterpiece,portrait,face to camera,**cowboy_hat**,smile

● 发带 hair_ribbon

提示词：4k,best quality,masterpiece,upper_body,face to camera,**hair_ribbon**,1 girl solo,casual wear,smile

● 发圈 hair_scrunchie

提示词：4k,best quality,masterpiece,upper_body,face to camera,**hair_scrunchie**,1 girl solo,smile

● 皇冠 crown

提示词：4k,best quality,masterpiece,upper_body,face to camera,**crown**,1 girl solo,smile

● 头冠 tiara

提示词：4k,best quality,masterpiece,upper_body,face to camera,**tiara**,1 girl solo,smile

● 环状耳环 hoop_earrings

提示词：4k,best quality,masterpiece,portrait,face to camera,**hoop_earrings**,1 girl solo,smile

● 十字耳环 cross_earrings

提示词：4k,best quality,masterpiece,portrait,face to camera,**cross_earrings**,1 girl solo,casual wear,smile

● 方形耳环 square_earrings

提示词：4k,best quality,masterpiece,portrait, face to camera,**square_earrings**,1 girl solo,smile

● 水晶耳环 crystal_earrings

提示词：4k,best quality,masterpiece,portrait, face to camera,**crystal_earrings**,1 girl solo,smile

● 花朵耳环 flower_earrings

提示词：4k,best quality,masterpiece,portrait, face to camera,**flower_earrings**,1 girl solo,smile

● 心形耳环 heart_earrings

提示词：4k,best quality,masterpiece,portrait, face to camera,**heart_earrings**,1 girl solo,smile

● 骷髅耳环 skull_earrings

提示词：4k,best quality,masterpiece,portrait,face to camera,**skull_earrings**,1 girl solo,casual wear,smile

● 星星耳环 star_earrings

提示词：4k,best quality,masterpiece,portrait,face to camera,**star_earrings**,1 girl solo,smile

● 民族风耳环 ethnic_earrings

提示词：4k,best quality,masterpiece,portrait,face to camera,**ethnic_earrings**,1 girl solo,smile

● 头巾 bandana

提示词：4k,best quality,masterpiece,portrait,face to camera,**bandana**,casual wear,smile

● 口罩　mouth_mask

提示词：4k,best quality,masterpiece,portrait, face to camera,**mouth_mask**

● 防尘口罩　dust_mask

提示词：4k,best quality,masterpiece,portrait, face to camera,**dust_mask**,1 girl solo

● 面部彩绘　facepaint

提示词：4k,best quality,masterpiece,portrait, face to camera,**facepaint**,smile

● 有色眼镜　tinted_eyewear

提示词：4k,best quality,masterpiece,portrait, face to camera,**tinted_eyewear**,smile

● 魅惑的微笑 seductive_smile

提示词：4k,best quality,masterpiece,upper_body,face to camera,**seductive_smile**,1 girl solo

● 大笑 laughing

提示词：4k,best quality,masterpiece,portrait,face to camera,**laughing**

● 咧嘴笑 grin

提示词：4k,best quality,masterpiece,upper_body,face to camera,**grin**

● 咯咯笑 giggling

提示词：4k,best quality,masterpiece,upper_body,face to camera,**giggling**,1 girl solo

伤心 sad

提示词：4k,best quality,masterpiece,portrait, face to camera,**sad**,1 girl solo

眉头紧锁 furrowed_brow

提示词：4k,best quality,masterpiece,portrait, face to camera,**furrowed_brow**

厌恶 disgust

提示词：4k,best quality,masterpiece,upper_ body,face to camera,**disgust**

蔑视 contempt

提示词：4k,best quality,masterpiece,upper_ body,face to camera,**contempt**,1 girl solo

● 害怕的 scared

提示词：4k,best quality,masterpiece,upper_body,face to camera,**scared**,1 girl solo

● 凝视 stare

提示词：4k,best quality,masterpiece,upper_body,face to camera,**stare**,1 girl solo

● 生气的 angry

提示词：4k,best quality,masterpiece,portrait,face to camera,**angry**

● 暴躁的 grumpy

提示词：4k,best quality,masterpiece,upper_body,face to camera,**grumpy**,male

● 不满的 discontented

提示词：4k,best quality,masterpiece,upper_body,face to camera,**discontented**,1 girl solo

● 困乏的 sleepy

提示词：4k,best quality,masterpiece,upper_body,face to camera,**sleepy**,1 girl solo

● 尖叫 screaming

提示词：4k,best quality,masterpiece,upper_body,face to camera,**screaming**

● 害羞的 shy

提示词：4k,best quality,masterpiece,upper_body,face to camera,**shy**,1 girl solo

Chapter 07

服装风格

服装风格是指人们在搭配和穿着服装时所追求的特定风格和时尚表达。不同的服装风格反映了个人的审美偏好、文化背景、时尚追求，以及社会群体的身份认同。

西装 suit

提示词：4k,best quality,masterpiece,full body,face to camera,**suit**,smile

水手服 sailor

提示词：4k,best quality,masterpiece,full body,face to camera,**sailor**,smile

● 校服 school_uniform

提示词：4k,best quality,masterpiece,full body,face to camera,**school_uniform**,smile

● 礼服裙 formal_dress

提示词：4k,best quality,masterpiece,full body,face to camera,**formal_dress**,1 girl solo,smile

和服式外套 kimono_jacket

提示词：4k,best quality,masterpiece,face to camera,**kimono_jacket**,1 girl solo,smile

婚纱 wedding_dress

提示词：4k,best quality,masterpiece,face to camera,**wedding_dress**,1 girl solo,smile

● **白衬衫** white_shirt

提示词：4k,best quality,masterpiece,upper_body,face to camera,**white_shirt**,smile

● **短袖T恤** T-shirts

提示词：4k,best quality,masterpiece,upper_body,face to camera, **T-shirts**,smile

● 夏威夷衫　hawaiian_shirt

提示词：4k,best quality,masterpiece,upper_body,face to camera,**hawaiian_shirt**,smile

● 连帽衫　hoodie

提示词：4k,best quality,masterpiece,face to camera,**hoodie**,smile

● 长袖运动卫衣 sweatshirt

提示词：4k,best quality,masterpiece,upper_body,face to camera,**sweatshirt**,smile

● 皮草大衣 fur_coat

提示词：4k,best quality,masterpiece,upper_body,face to camera,**fur_coat**,smile

派克大衣　parka

提示词：4k,best quality,masterpiece,upper_body,face to camera,**parka**,smile

夹克衫　jacket

提示词：4k,best quality,masterpiece,upper_body,face to camera,**jacket**,smile

● 蓬蓬裙　bubble_skirt

提示词：4k,best quality,masterpiece,face to camera,**bubble_skirt**,1 girl solo,smile

● 中筒袜　kneehighs

提示词：4k,best quality,masterpiece,full body,face to camera,**kneehighs**,1 girl solo,smile

高跟鞋　high_heels

提示词：4k,best quality,masterpiece,full body,face to camera,**high_heels**,1 girl solo,smile

雨靴　rain_boots

提示词：4k,best quality,masterpiece,full body,face to camera,**rain_boots**,1 girl solo,smile

● 紧身连衣裤　leotard

提示词：4k,best quality,masterpiece,full body,face to camera,**leotard**,smile

● 套装　costume

提示词：4k,best quality,masterpiece,face to camera,**costume**,smile

连帽斗篷 hooded_cloak

提示词：4k,best quality,masterpiece,upper_body,face to camera,**hooded_cloak**,smile

冬装 winter_clothes

提示词：4k,best quality,masterpiece,upper_body,face to camera,**winter_clothes**,smile

● **工装裤** cargo_pants

提示词：4k,best quality,masterpiece,full body,face to camera,**cargo_pants**,smile

● **健身短裤** gym_shorts

提示词：4k,best quality,masterpiece,face to camera,**gym_shorts**,1 girl solo,smile

厨师制服   chef_uniform

提示词：4k,best quality,masterpiece,face to camera,**chef_uniform**,smile

实验服   labcoat

提示词：4k,best quality,masterpiece,face to camera,**labcoat**,smile

● 警察制服　police_uniform

提示词：4k,best quality,masterpiece,face to camera,**police_uniform**,smile

● 棒球外套　letterman_jacket

提示词：4k,best quality,masterpiece,upper_body,face to camera,**letterman_jacket**,smile

排球服　volleyball_uniform

提示词：4k,best quality,masterpiece,face to camera,**volleyball_uniform**,smile

摔跤服　wrestling_outfit

提示词：4k,best quality,masterpiece,face to camera,**wrestling_outfit**,smile

● 拉拉队服 cheerleader uniform

提示词：4k,best quality,masterpiece,upper_body,face to camera,**cheerleader uniform**,1 girl solo,smile

● 泳装 swimsuit

提示词：4k,best quality,masterpiece,face to camera,**swimsuit**,smile

● 骑行短裤 bike_shorts

提示词：4k,best quality,masterpiece,face to camera,**bike_shorts**,1 girl solo,smile

● 围裙 apron

提示词：4k,best quality,masterpiece,face to camera,**apron**,1 girl solo,smile

● 领带 necktie

提示词：4k,best quality,masterpiece,upper_body,face to camera,**necktie**,male,smile

● 条纹围巾 striped_scarf

提示词：4k,best quality,masterpiece,upper_body,face to camera,**striped_scarf**,smile

哥特风格　gothic

提示词：4k,best quality,masterpiece,face to camera,**gothic**,smile

哥特洛丽塔　gothic_lolita

提示词：4k,best quality,masterpiece,upper_body,face to camera,**gothic_lolita**,1 girl solo,smile

● 公主裙　princess_dress

提示词：4k,best quality,masterpiece,face to camera,**princess_dress**,1 girl solo,smile

● 西部风格　western

提示词：4k,best quality,masterpiece,face to camera,**western**,smile

女巫装   witch_clothes

提示词：4k,best quality,masterpiece,face to camera,1girl solo,**witch_clothes**,smile

修女装   nun_clothes

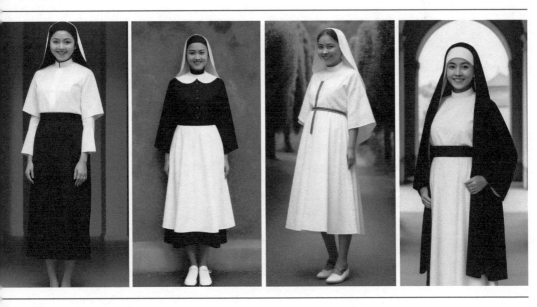

提示词：4k,best quality,masterpiece,face to camera,1girl solo,**nun_clothes**,smile

● 芭蕾服　ballerina

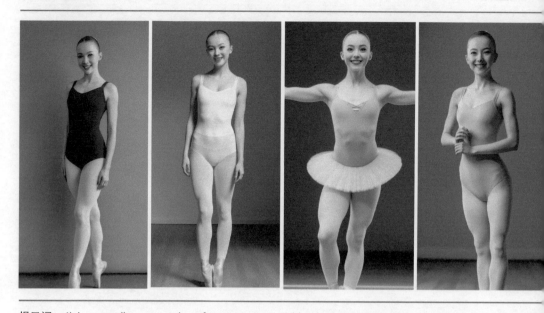

提示词：4k,best quality,masterpiece,face to camera,1girl solo,**ballerina**,smile

● 演出服　singing costume

提示词：4k,best quality,masterpiece,face to camera,1girl solo,**singing costume**,smile

圣诞风格服装　santa_costume

提示词：4k,best quality,masterpiece,face to camera,**santa_costume**,smile

万圣节服装　halloween_costume

提示词：4k,best quality,masterpiece,face to camera,**halloween_costume**,smile

Chapter 08

人物动作

人物动作是指人们在日常生活、表演或其他情境中所展现的身体动作和姿态。人物动作可以用来表达情感、传达信息、展示个性，以及推动剧情或故事的发展。本章展示了一些常见的人物动作。

● 站立 standing

提示词：4k,best quality,masterpiece,full body,face to camera,**standing**,smile

● 身体往后靠 leaning_back

提示词：4k,best quality,masterpiece,face to camera,**leaning_back**,smile

● 靠在一边 leaning_to_the_side

提示词：4k,best quality,masterpiece,full body,face to camera,**leaning_to_the_side**,smile

● 奔跑 running

提示词：4k,best quality,masterpiece,full body,face to camera,**running**,smile

身体倾斜　body_leaning

提示词：4k,best quality,masterpiece,face to camera,**body_leaning**,smile

摆姿势　posing

提示词：4k,best quality,masterpiece,face to camera,**posing**,smile

● **躺着** lying_down

提示词：4k,best quality,masterpiece,upper_body,face to camera,**lying_down**,1 girl solo,smile

● **追逐** chasing

提示词：4k,best quality,masterpiece,full body,face to camera,**chasing**,smile

双臂交叉 arms crossed

提示词：4k,best quality,masterpiece,full body,face to camera,**arms crossed**,male,smile

招手 waving

提示词：4k,best quality,masterpiece,face to camera,**waving**,smile

● 打扫 cleaning

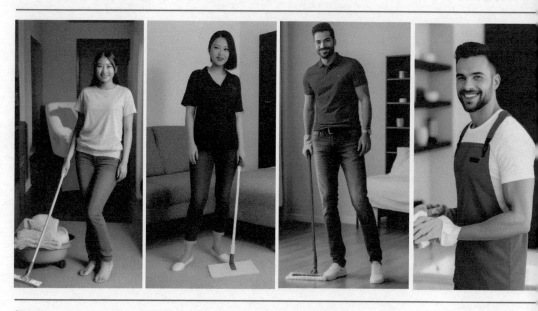

提示词：4k,best quality,masterpiece,face to camera,**cleaning**,smile

● 攀爬 climbing

提示词：4k,best quality,masterpiece,full body,face to camera,**climbing**,smile

祈祷 praying

提示词：4k,best quality,masterpiece,face to camera,**praying**,1 girl solo,smile

跪 kneeling

提示词：4k,best quality,masterpiece,full body,face to camera,**kneeling**,smile

● 半蹲 half squat

提示词：4k,best quality,masterpiece,full body,face to camera,**half squat**,smile

● 挥拳 punching

提示词：4k,best quality,masterpiece,face to camera,**punching**,smile

骑行 riding

提示词：4k,best quality,masterpiece,full body,face to camera,**riding**,smile

唱歌 sing

提示词：4k,best quality,masterpiece,full body,face to camera,**sing**,smile

● 游泳 swimming

提示词：4k,best quality,masterpiece,face to camera,**swimming**,smile

● 拿着手机 holding_phone

提示词：4k,best quality,masterpiece,face to camera,**holding_phone**,smile

提着包 holding_bag

提示词：4k,best quality,masterpiece,face to camera,**holding_bag**,1 girl solo,smile

拿着球 holding_ball

提示词：4k,best quality,masterpiece,face to camera,**holding_ball**,smile

● 拿着书　holding_book

提示词：4k,best quality,masterpiece,face to camera,**holding_book**,smile

● 拎着篮子　holding_basket

提示词：4k,best quality,masterpiece,face to camera,**holding_basket**,smile

● 拿着花束 holding_bouquet

提示词：4k,best quality,masterpiece,face to camera,**holding_bouquet**,smile

● 拿着碗 holding_bowl

提示词：4k,best quality,masterpiece,face to camera,**holding_bowl**,1 girl solo,smile

● 拿着玩偶 holding_doll

提示词：4k,best quality,masterpiece,upper_body,face to camera,**holding_doll**,1 girl solo,smile

● 拿着礼物 holding_gift

提示词：4k,best quality,masterpiece,face to camera,**holding_gift**,smile

捋头发 hair_tucking

提示词：4k,best quality,masterpiece,face to camera,**hair_tucking**,1 girl solo,smile

拥抱 hug

提示词：4k,best quality,masterpiece,full body,face to camera,**hug**,smile

Chapter 09

场景呈现

场景呈现是指在电影、戏剧等艺术作品的创作中,通过布景、灯光和道具等手段,将虚拟或现实的场景呈现给观众的过程。通过场景的精心设计,可以为故事情节提供背景环境、营造氛围,增强观众的情感体验和沉浸感。本章给出了一些常见场景呈现的要素。

● 教室 classroom

提示词:4k,best quality,masterpiece, **classroom**

● 咖啡厅 café

提示词:4k,best quality,masterpiece,**café**

● 植物园 botanical_garden

提示词：4k,best quality,masterpiece, **botanical_garden**

● 足球场 soccer_field

提示词：4k,best quality,masterpiece, **soccer_field**

游乐园 amusement_park

提示词：4k,best quality,masterpiece,**amusement_park**

居酒屋 Izakaya

提示词：4k,best quality,masterpiece,**Izakaya**

● 宴会 banquet

提示词：4k,best quality,masterpiece, **banquet**

● 电车内 train_interior

提示词：4k,best quality,masterpiece,**train_interior**

游泳池 swimming pool

提示词：4k,best quality,masterpiece, **swimming pool**

办公室 office

提示词：4k,best quality,masterpiece,**office**

● 卧室 bedroom

提示词：4k,best quality,masterpiece, **bedroom**

● 厨房 kitchen

提示词：4k,best quality,masterpiece,**kitchen**

中式阁楼　chinese_style_loft

提示词：4k,best quality,masterpiece,**chinese_style_loft**

礼堂　auditorium

提示词：4k,best quality,masterpiece,**auditorium**

● **汽车座椅** car_seat

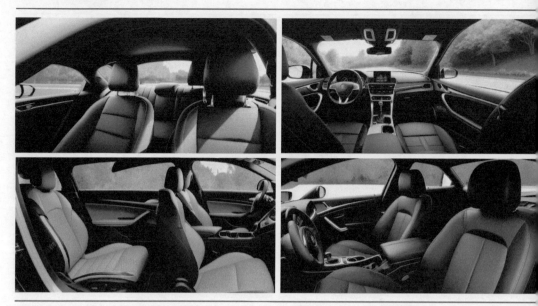

提示词：4k,best quality,masterpiece, **car_seat**

● **洗衣店** laundry

提示词：4k,best quality,masterpiece,**laundry**

● 实验室 laboratory

提示词：4k,best quality,masterpiece,**laboratory**

● 书店 bookstore

提示词：4k,best quality,masterpiece,**bookstore**

● 演唱会 concert

提示词：4k,best quality,masterpiece,**concert**

● 自动贩卖机 vending_machine

提示词：4k,best quality,masterpiece,**vending_machine**

● 街景 street scenery

提示词：4k,best quality,masterpiece,**street scenery**

● 雨天街景 rain street scenery

提示词：4k,best quality,masterpiece,**rain street scenery**

● 草原 grasslands

提示词：4k,best quality,masterpiece, **grasslands**

● 花海 flower_field

提示词：4k,best quality,masterpiece,**flower_field**

雪山 snowy_mountain

提示词：4k,best quality,masterpiece, **snowy_mountain**

星空 starry_sky

提示词：4k,best quality,masterpiece,**starry_sky**

● 小岛　floating_island

提示词：4k,best quality,masterpiece, **floating_island**

● 森林　forest

提示词：4k,best quality,masterpiece,**forest**

东亚建筑 east_asian_architecture

提示词：4k,best quality,masterpiece, **east_asian_architecture**

鸟居 torii

提示词：4k,best quality,masterpiece,**torii**

● **大教堂** cathedral

提示词：4k,best quality,masterpiece,**cathedral**

● **摩天大楼** skyscraper

提示词：4k,best quality,masterpiece,**skyscraper**

城堡 castle

提示词：4k,best quality,masterpiece, **castle**

清真寺 mosque

提示词：4k,best quality,masterpiece,**mosque**

● 铁路 railroad

提示词：4k,best quality,masterpiece,**railroad**

● 桥 bridge

提示词：4k,best quality,masterpiece,**bridge**

废墟 ruins

提示词：4k,best quality,masterpiece, **ruins**

海滩 beach

提示词：4k,best quality,masterpiece,**beach**

● 春天 spring

提示词：4k,best quality,masterpiece,**spring**

● 夏天 summer

提示词：4k,best quality,masterpiece,**summer**

秋天 autumn

提示词：4k,best quality,masterpiece,**autumn**

冬天 winter

提示词：4k,best quality,masterpiece,**winter**

● 火山 volcano

提示词：4k,best quality,masterpiece, **volcano**

● 悬崖 cliff

提示词：4k,best quality,masterpiece,**cliff**

● 赛博朋克 cyberpunk

提示词：4k,best quality,masterpiece, **cyberpunk**

● 幻想风格 fantasy

提示词：4k,best quality,masterpiece,**fantasy**

● **万圣节** halloween

提示词：4k,best quality,masterpiece, **halloween**

● **圣诞节** christmas

提示词：4k,best quality,masterpiece,**christmas**